Australian general practitioner perceptions of the detection and screening of at-risk drinking, and the role of the AUDIT-C

Research Thesis as partial requirement of:
Master of Mental Health (General Practitioner) (Research Pathway)

The New South Wales Institute of Psychiatry

Chun Wah Michael Tam
BSc(Med) MBBS FRACGP

Submitted: 28 December 2012

The New South Wales Institute of Psychiatry

SCHOOL OF PUBLIC HEALTH AND COMMUNITY MEDICINE

Declaration

This thesis is my own work and no part of it has been submitted for a degree at this, or any other university.

Signed:

Michael Tam

Date: 28 December 2012

Acknowledgements

There are many people that I must thank for their involvement and support in this research.

Firstly, I would like to express my gratitude to Dr Roslyn Markham for her extensive teaching, support and supervision throughout this long process. She assisted my learning in research methodology, helped guide and develop my research question, and provided invaluable feedback to the manuscript.

Secondly, my warm appreciations go to Prof Nicholas Zwar for his experienced guidance and mentorship. Without his support and offer to supervise my research in general practice, this project could not have taken place.

To Natalie Healy; for her organisation and education support at the NSW Institute of Psychiatry.

To Oshana Hermiz at the Centre for Primary Health Care and Equity; for his assistance during the early phase of the focus groups.

To Melanie Marshall at the UNSW; for organisation support with funding for transcription.

To Dr Joel Rhee at the UNSW; for his assistance with using Nvivo 9.

To my partner, May Su; for her enduring patience and understanding.

I also thank all the general practitioners who agreed to participate in this research project.

Finally, I acknowledge the NSW Health Department who supported this research through a scholarship for the Master of Mental Health (General Practitioner).

Preface

The first glimmer of the idea that eventually led to this research project was in late 2009. I was enrolled in the "Mental Illness and Substance Abuse" course at the New South Wales Institute of Psychiatry. In one of the assessment tasks, I used the Alcohol Use Disorders Identification Test as an opportunistic screening tool on five patients in a general practice setting.

In 2011, I stepped side-wards into a part-time academic position in the School of Public Health and Community Medicine at the University of New South Wales. This opportunity enabled me to reflect more thoroughly on my prior experiences.

In my discussions with my research supervisors (Dr Roslyn Markham and Prof Nicholas Zwar) and early reading of the literature, I was struck by the apparent incongruity between clinical recommendations and clinical practice. In the field of detection and treatment of alcohol use disorders, there is voluminous research on a variety of questionnaire based screening tools, with some recognition that they are not used by general practitioners (GPs). Most of the recent research was European and American, and some of the findings did not resonate with me as an Australian GP. Moreover, despite the primary care focus, little of the research was conducted by GPs.

A common theme that appears in the literature is that the low detection of at-risk drinking is due to deficiencies in GP knowledge and attitudes. An egregious example from 1998 speaks of GPs' "negative attitudes", "lack of training", "a profession lacking in confidence", "low levels of therapeutic commitment", and "failure to meet the 'Health of the Nation' [UK Government alcohol policy targets]" (Deehan, Templeton, Taylor, Drummond, & Strang, 1998) – and this only in the title and abstract! This theme is reflected by an assumption that appears frequently, that more GP education and support on the use of screening and brief interventions is needed and necessarily beneficial. I wondered about this assumption when I reflected on my own experience. Despite receiving post-graduate education and training in alcohol screening questionnaires, I did not once feel inclined to use one of these tools after the assignment.

This project is my first foray into primary care research. I wanted to develop and share a deeper understanding into this apparent gap between research and practice in the general practice setting.

Part II contains the article that was submitted to *Drug and Alcohol Review*.

Table of Contents

Abstract

At-risk drinking is common in Australia. Validated screening tools such as the AUDIT-C have been promoted to general practitioners (GPs), but appear rarely used. Detection of at-risk drinking remains low.

In this research, we used qualitative methods to describe Australian GP perceptions of the detection and screening of at-risk drinking. We sought to understand the low uptake of alcohol screening questionnaires in general practice, the potential role of the AUDIT-C, and the overall low detection rate of at-risk drinking.

A convenience sample of GP teaching practices in metropolitan Sydney were approached to participate in the study. Four focus group interviews of GPs and GP trainees were held, with a total of 19 participants. These were conducted between August and October 2011. The audio recordings of these interviews were transcribed and analysed using Straussian grounded theory methodology.

Four major factors arose in the results: (i) GP perceptions of the detection of at-risk drinking, (ii) sociocultural attitudes towards drinking, (iii) dynamics of the patient-doctor relationship, and (iv) GP perceptions of alcohol screening and the AUDIT-C.

Sociocultural factors appear to have a key influence on the major barriers to detection of at-risk drinking in primary care. These barriers were: community stigma and stereotypes of "problem" drinking, GP perceptions of unreliable patient alcohol use histories, and the perceived threat to the patient-doctor relationship from alcohol use assessment. Alcohol screening questionnaires are not designed to address these factors and barriers. The AUDIT-C in particular was seen to have poor practical utility.

In the current context, it is unlikely that approaches that focus on the use of these tools will be effective at improving detection of at-risk drinking by GPs.

Part I

Literature Review

Part I – Literature Review

1.1 Introduction

"At-risk drinking" refers to the consumption of alcohol in such a manner that an individual is placed at increased risk of alcohol-related harm. This manner of drinking is common in Australia (Australian Institute of Health and Welfare [AIHW], 2008, 2011; Britt et al., 2010a, 2010b; National Health and Medical Research Council [NHMRC], 2009). GPs in primary care settings are seen to have a major role in reducing these harms – they are often an individual's first point of contact with the health system and have good access to the at-risk drinking population (Ministerial Council on Drug Strategy, 2006). Patients expect preventive and health promotion messages from GPs (Aalto, Pekuri, & Seppa, 2002; Richmond, Kehoe, Heather, Wodak, & Webster, 1996; P. G. Wallace et al., 1984, 1987). Furthermore, brief alcohol interventions appear to be effective when delivered in primary care settings in reducing patient alcohol intake (Kaner et al., 2009).

However, for GPs to offer and provide brief alcohol interventions, they must first detect at-risking drinking in their patients. Alcohol screening questionnaires have been broadly promoted for this purpose and GPs have been told to use them. But, uptake of these tools is low (Aalto, Pekuri, & Seppa, 2003; Engdahl & Nilsen, 2011; Friedmann, McCullough, Chin, & Saitz, 2000; Nygaard, Paschall, Aasland, & Lund, 2010; Spandorfer, Israel, & Turner, 1999). Recent evidence suggests that GPs across many countries do not identify the majority of at-risk drinkers and brief interventions are rarely offered (Berner et al., 2007a; Britt et al., 2010a; D'Amico, Paddock, Burnam, &

Kung, 2005; Solberg, Maciosek, & Edwards, 2008; Yamada, Maeno, Waza, & Sato, 2008).

At present, there is no coherent consensus on either the cause or solution to this problem.

1.2 At-risk drinking

One difficulty for the field as a whole is the lack of an internationally accepted and consistent nomenclature. The terms "risky drinking" and "hazardous drinking" are often used synonymously with at-risk drinking. At times, the terms "harmful drinking", "excessive drinking", "heavy drinking" and "problem drinking" are used in a similar manner, though these expressions refer more to patterns of drinking that are directly harmful to health (rather than simply increasing the risk of harm). This ambiguity in terminology is reflected in that authors of contemporary articles discussing at-risk drinking typically still explicitly define the usage of the term within their paper (Bradley, Kivlahan, & Williams, 2009).

Although there is general agreement on the *conceptual* meaning of at-risk drinking (a pattern of drinking that places an individual at increased risk of alcohol-related harm), how it has been *operationally* defined varies considerably. As assessment of patterns of drinking includes the quantity of alcohol consumed, the first point of confusion is the commonly used concept of the "standard drink". Again, there is no internationally consistent definition. The Australian standard drink containing 10 grams of ethanol (Department of Health and Ageing, 2009; Russell et al., 1994), but the UK standard drink is only 8 grams (Science and Technology Committee, 2012) while the US standard drink is 14 grams (Centers for Disease Control and Prevention, 2012). Care must be taken when interpreting non-Australian literature!

Within the published literature that is pertinent to Australian primary care, at-risk drinking has been defined in three main ways: (i) the National Health and Medical Research Council (NHMRC) 2001 guidelines definition, (ii) the definition based on the NHMRC 2009 guidelines, and (iii) the Bettering the Evaluation and Care of Health (BEACH) project definition. These three definitions demonstrate different conceptual approaches to alcohol risks and harms.

The first definition from the former (and rescinded) NHMRC (2001) alcohol drinking guidelines is particularly important as it was widely used in the Australian context. These guidelines conceptualised that risk could be both short-term (i.e., the harm associated with drinking on a given day), and long-term (i.e., the harm associated with regular patterns of drinking). Moreover, it described these risks using consumption thresholds (Ibid). For instance, at-risk drinking for men was consuming more than 6 standard drinks on a given day, or more than 4 standard drinks per day on average. Similarly, at-risk drinking for women was consuming more than 4 standard drinks on a given day, or more than 2 standard drinks per day on average. Interestingly, contemporary community perceptions of short-term risks of harms from alcohol appear to remain closely linked to these older guidelines (NHMRC, 2009, p. 17).

One of the major conceptual changes in the current NHMRC (2009) guidelines is that it no longer specifies threshold "risky" levels of drinking. Indeed, it explicitly avoids the term "risky drinking" as the authors considered it "hard to quantify" (Ibid, p. 149). It considers risk of harm as progressively increasing with higher alcohol consumption.

6

Instead, it describes levels of drinking that are "low risk". In the current guidelines, low risk drinking for both men and women is no more than 4 standard drinks on a single occasion, and no more than 2 standard drinks per day over a lifetime (Ibid).

This conceptual shift in the current NHMRC (2009) guidelines makes it ambiguous as to what consumption level is at-risk drinking. Simply put, should drinking in excess of *low risk* recommendations be categorically considered to be *at-risk* drinking? The NHMRC (2009) guidelines itself does not answer this question. Nevertheless, at-risk drinking thresholds are broadly used in alcohol research. The Australian Institute of Health and Welfare (AIHW) answered the above question in the affirmative in the 2010 National Drug Strategy Household Survey (NDSHS). Based on the NHMRC (2009) guidelines, it defined at-risk drinking in adults to be drinking more than 4 standard drinks on a single occasion, or more than 2 standard drinks a day on average (AIHW, 2011).

Notably, this interpretation of the NHMRC (2009) guidelines represents a practical lowering of the at-risk drinking thresholds for men compared to the 2001 guidelines – from 6 to 4 standard drinks on a given day/occasion of drinking, and from 4 to 2 standard drinks per day on average.

The third definition of at-risk drinking of importance in the Australian setting comes from the ongoing BEACH project. This research initiative has been gathering patient visit information in Australian general practice since 1998 and the dataset now includes details of approximately 1.4 million GP consultations (Family Medicine Research Centre, 2012). The BEACH project uses the AUDIT-C questionnaire (Table 1) to identify at-risk drinkers (Bradley et al., 2007; Bush et al., 1998). BEACH defines men with an AUDIT-C score of 5 or greater, and women with a score of 4 or greater as at-risk drinkers (Britt et al., 2010a, p. 134).

This definition is conceptually different again from the current and former NHMRC definitions. The use of an alcohol screening questionnaire is a more clinical approach to identifying at-risk drinkers – this is consistent with the method of data collection (GPs completing forms on consecutive patient presentations). Furthermore, the definition by score thresholds is potentially at odds with the definition based on NHMRC (2009) guidelines. As an example, a woman who drinks 4 or more times a week, would receive an AUDIT-C score of at least 4 and thus be classified as an at-risk drinker in the BEACH data. However, she would only be considered an at-risk drinker using the NHMRC (2009) definition if she drank more than 2 standard drinks per day.

1.2.1 Prevalence and health burden

Interestingly, despite the different operational definitions of at-risk drinking, the empirically measured prevalence rates using them appear largely consistent. The most recent Australian community data available is from the 2010 NDSHS. This survey examined the use of drugs by Australians in the community, as well as their perceptions and attitudes to drugs (AIHW, 2011). The prevalence of at-risk

Table 1 AUDIT-C questionnaire

	Score
How often do you have a drink containing alcohol?	
Never	+0
Monthly or less	+1
2-4 times a month	+2
2-3 times a week	+3
4 or more times a week	+4
How many standard drinks containing alcohol do you have on a typical day?	
1 or 2	+0
3 or 4	+1
5 or 6	+2
7 or 9	+3
10 or more	+4
How often do you have six or more drinks on one occasion?	
Never	+0
Less than monthly	+1
Monthly	+2
Weekly	+3
Daily or almost daily	+4

Adapted from Harris et al. (2009a)

drinking based on the current NHMRC (2009) guidelines was (AIHW, 2011, Table 4.4, Table 4.5, Table 4.6, pp. 52-53, 55-57, 59):

- Increased single occasion risk (at least weekly):   15.9%

- Increased lifetime risk: 20.1%

- **Increased single occasion OR lifetime risk: 22.4%**

Prior to the 2010 NDSHS, the AIHW used the NHMRC (2001) guidelines risk definitions (AIHW, 2008). In 2007, the prevalence of at-risk drinking using this older guideline was 22.1% (AIHW, 2008, Table 5.2, p. 32). When this older data is analysed using the current NHMRC guidelines, the prevalence of at-risk drinking in 2007 is a very similar 22.7% (calculated by the author) (AIHW, 2011, Table 4.5, pp. 55-57). Appendix 2.1 describes the assumptions used in this calculation in greater detail.

The BEACH project provides data on patients in general practice, rather than of the general community. In the most recent publicly available data from Britt et al. (2010a), 26.5% of patients were classified as at-risk drinkers (Table 14.3, p. 135). When the data is weighted to adjust for age and sex attendance patterns (women and older people are over-represented in the BEACH sample), the at-risk drinking prevalence for the general practice patient population rises to 29.7%. This is not overly different to the aforementioned community prevalence rate (22.4%) from the 2010 NDSHS. It is expected that the prevalence rate would be higher in patients than in the community sample as at-risk drinkers are more likely to see GPs for management of a range of chronic health conditions (Proude, Britt, Valenti, & Conigrave, 2006). Other findings in the BEACH data were that younger people had a higher prevalence of at-risk drinking than older individuals, and that men had a higher prevalence than women in all age groups. These patterns are consistent with general community data (AIHW, 2011).

Although the current at-risk drinking prevalence is known with some precision, the direction of its trend is less clear. There was reported to be a decline in the proportion of Australians who consumed alcohol daily between 2001 and 2010, from 8.3% to 7.2% (AIHW, 2011, Table 4.1, p. 46). However, there did not appear to have been a commensurate reduction in the prevalence of at-risk drinking over the same period (Ibid, Figure 4.2, p. 58). The BEACH data suggests that the prevalence of at-risk drinking in the patient population may have increased slightly over the past decade – though this change did not reach statistical significance (Britt et al., 2010b, Table 14.1, p. 126). As a further complication, the more recent figures of per capita consumption of alcohol may have been underestimates. Chikritzhs, Allsop, Moodie and Hall (2010) report that there has been a hidden increase in alcohol consumption in the past two decades due to a gradual increase in the alcohol content of wine.

Whatever the real trend is of at-risk drinking, it is notably different to the trend in tobacco consumption, the other commonly used substance in the community. Tobacco use has seen a major change in prevalence over the same period, an overall 20% reduction (AIHW, 2011, Table 3.1, p. 23). In comparison, at-risk drinking rates are probably unchanged, and possibly higher. These substance use trends suggest that alcohol is an increasingly important contributor to the burden of disease and injury in Australia. Behind tobacco, it is already the second greatest preventable cause of drug-related deaths and hospitalisation (NHMRC, 2009) and in 2003 alcohol use was estimated to have contributed to 3.2% of the total burden of disease and injury (Begg, Vos, Barker, Stevenson, & Lopez, 2007, p. 84).

Part I – Literature Review

1.3 The general practitioner's formal role

At-risk drinking is common and alcohol consumption is a significant preventable cause of disease and injury. This recognition led the World Health Organization (WHO) in the early 1980s to advocate a preventive and health promotion focussed approach to alcohol-related harms (J. B. Saunders, Aasland, Babor, Delafuente, & Grant, 1993b). Primary care was identified as a setting with enormous potential – GPs have access to the population of at-risk drinkers, often before the occurrence of alcohol-related harm. The WHO's goals were for at-risk drinkers to be identified early in the primary care setting, for brief interventions to be delivered to those who might benefit from them, and thus lessen the overall burden of alcohol-related disease and illness in populations (Ibid).

This approach has been subsequently supported by (i) the development of validated screening questionnaires (Berner, Kriston, Bentele, & Harter, 2007b; Meneses-Gaya, Zuardi, Loureiro, & Crippa, 2009; Reinert & Allen, 2007), (ii) evidence of the effectiveness of brief alcohol interventions (Kaner et al., 2009), and (iii) economic modelling suggestive of its cost-effectiveness compared to other GP health promotion activities (Solberg et al., 2008). As a strategy to reduce the health impacts of alcohol consumption, it has been implemented as policy in Australia at multiple levels.

The Australian Government's National Alcohol Strategy 2006-2011 identifies GPs as a key health practitioner in the priority area of "health impacts". It mandates that the health system should "promote primary care settings as an accessible and non-stigmatising opportunity for health promotion, prevention and treatment of alcohol use problems" (Ministerial Council on Drug Strategy, 2006, p. 25).

Similarly, the peak representative body of GPs, the Royal Australian College of General Practitioners (RACGP) support the principles of early detection and intervention in two clinical practice guidelines (Harris et al., 2009b; RACGP, 2004). In terms of clinical activity, the RACGP recommends that (i) GPs should routinely and regularly screen their patients for at-risk drinking from adolescence; (ii) GPs should consider using the Alcohol Use Disorders Identification Test (AUDIT) (RACGP, 2004, p. 20) or AUDIT-C (Harris et al., 2009b, p. 39) screening questionnaires; and (iii) GPs should offer brief alcohol interventions to patients with at-risk drinking.

The adoption of this approach to managing alcohol problems, and the formalisation of the role of GPs in Australia are consistent with the policy actions seen in other countries. This is perhaps best reflected by the concordance of the RACGP's alcohol recommendations with those of the US Preventive Services Task Force (2004), and National Institute for Health and Clinical Excellence (2011) in the UK. All three countries recommend routine screening for at-risk drinking using screening questionnaires, and offering brief interventions to at-risk drinkers.

However, despite their formalised role in managing alcohol problems, it does not appear that Australian GPs have embraced the initial steps of the health promotion approach – the identification of at-risk drinkers by routine screening.

Early evidence from the 1980s and 1990s demonstrated that Australian GPs identified no more than a third of at-risk drinkers (Reid, Webb, Hennrikus, Fahey, & Sanson-Fisher, 1986; Rydon, Redman, Sanson-Fisher, & Reid, 1992). These findings are consistent with international observations from that period (Buchsbaum, Buchanan, Poses, Schnoll, & Lawton, 1992; Leckman, Umland, & Blay, 1984; Vande Creek, Zachrich, & Scherger, 1982).

More recently, a large primary care study in 2007 demonstrated that GPs in Germany detected only 33.6% of at-risk drinkers (Berner et al., 2007a). Indirect evidence from a 2002 patient survey in Finland suggests a similar detection rate – between 19.0 to 47.6% (Aalto et al., 2002). Lastly, there is a curious small study of a solo practitioner family practice in Japan (Yamada et al., 2008). The GP in this study, who had an interest in mental health, surprisingly identified none of the at-risk drinkers out of a series of consecutive first-visit patients!

Importantly, it appears that contemporary GP detection rates of at-risk drinking are not meaningfully different to historical ones – the majority of at-risk drinkers continue to be not identified. Moreover, the delivery of alcohol counselling in Australian general practice has remained static at the low rate of 4 in 1000 consultations over the last decade (Britt et al., 2010b, Table 10.1b, p. 91).

Part I – Literature Review

1.4 Alcohol screening questionnaires

In recognition of the low detection rates of at-risk drinking, alcohol screening questionnaires have been widely promoted as an effective tool. Since the late 1960s, there have been many alcohol screening questionnaires that have been developed, studied, and validated (see Appendix 2.2 for a non-exhaustive list). The three major independent questionnaires, the CAGE, the Michigan Alcoholism Screening Test (MAST), and the AUDIT, will be described in detail. Most other questionnaires use items from one or more of these instruments.

The CAGE and MAST are the forerunners of all other questionnaires and their construction need to be considered within the historical context of the 1960s and 1970s. This period saw a major theoretical shift in the field of psychiatry – from a foundation based on psychodynamic theory, to an "empirical, positivistic orientation" (Sanders, 2011). The reasons for this change were complex, but include the unreliability and subjectivity of psychiatric diagnoses, and the desire to return to psychiatry's "medical roots" (Ibid, p. 398). The developers of the CAGE and MAST, Ewing and Selzer respectively, both noted that the lack of a consistent definition of "alcoholism" was the rationale for the development of a tool in aiding detection and diagnosis (Ewing, 1984; Selzer, 1971). Selzer opined, "most of the definitions are couched in terms so broad that their diagnostic usefulness is often limited to patients who are grossly alcoholic" (Selzer, 1971, p. 89).

The major limitations to both the CAGE and MAST are that they were designed for the detection of "alcoholism" and were tested in hospital patients (Ewing, 1984; Selzer, 1971). They are not as suited for the detection of milder alcohol drinking problems, especially in primary care populations. Notably, they did not detect the majority of at-risk drinkers when tested in community patients (W. M. Saunders & Kershaw, 1980).

In the context of the WHO's preventive and health promotion approach to alcohol problems, consensus amongst researchers in the early 1980s was that a new screening instrument was needed, one specifically designed for use in primary care (J. B. Saunders et al., 1993b). The CAGE and MAST were considered inadequate. Biochemical markers were determined to lack sufficient sensitivity for screening purposes (Aertgeerts, Buntinx, Ansoms, & Fevery, 2001; Bernadt, Mumford, Taylor, Smith, & Murray, 1982; J. B. Saunders et al., 1993b). A multinational WHO collaborative project was formed and this led to the creation of the AUDIT questionnaire, first published in 1989 (J. B. Saunders et al., 1993a, 1993b).

The AUDIT has since been extensively tested and validated in a wide range of international populations and clinical settings. Two recent systematic reviews of the AUDIT concluded that it is psychometrically sound with better performance than other screening methods (Meneses-Gaya et al., 2009; Reinert & Allen, 2007), and both recommended its broad application. A third systematic review suggested a slightly more cautious approach, recommending that it should only be used in populations similar to those on which it had been validated (Berner et al., 2007b).

The apparent clinical effectiveness (Berner et al., 2007b; Meneses-Gaya et al., 2009; Reinert & Allen, 2007), and cost-effectiveness (Solberg et al., 2008) of alcohol screening questionnaires, has led alcohol researchers to advocate the use of these tools in general practice as a necessary component in the strategy of reducing alcohol-related harm.   However, it appears that alcohol screening questionnaires are used by few GPs (Aalto et al., 2003; Engdahl & Nilsen, 2011; Friedmann et al., 2000; Nygaard et al., 2010; Spandorfer et al., 1999). Even when a screening instrument is used, many GPs choose the older CAGE in lieu of the more psychometrically sound AUDIT (Spandorfer et al., 1999).

From the literature, it appears that alcohol researchers have responded to the phenomenon of poor uptake of screening questionnaires in two ways main ways. The first were the attempts to simplify the questionnaires. For instance, the 10 items of the AUDIT proved unwieldy in routine primary care (Beich, Gannik, & Malterud, 2002; Berner et al., 2007b; Bradley et al., 2007; Bradley et al., 2009) and a variety of shortened versions have been subsequently developed and validated (Meneses-Gaya et al., 2009; Reinert & Allen, 2007). Of these, the AUDIT-C and FAST remain in contemporary use (Bradley et al., 2007; Demirkol, Haber, & Conigrave, 2011; Meneses-Gaya et al., 2010). A number of authors have even suggested the use of single question screening tools (Bradley et al., 2009; Seale et al., 2006).

The second (and predominant) response has been the focus on GP-related factors such as their knowledge, skills and confidence. Various authors have postulated that if only GPs had better training, were more confident, and were supported in the use of alcohol screening tools and brief interventions, that uptake of these preventive and health promotional activities would be improved (Berner et al., 2007a; Deehan et al., 1998; Durand, 1994; Kaner et al., 2009; Reid et al., 1986; Rydon et al., 1992; P. Wallace & Haines, 1985).

There is some question as to the effectiveness and rationale of these two responses. In regards to the first, the original CAGE questionnaire had only 4 items with a simple scoring scheme. Nevertheless, the historical uptake of this tool was limited (Nygaard et al., 2010; Spandorfer et al., 1999). Furthermore, there is no evidence that suggests increased uptake of simpler versions of the AUDIT (Nygaard et al., 2010), or that patients have been asked about their use of alcohol more frequently since their introduction (Britt et al., 2010b; Engdahl & Nilsen, 2011). With regards to the second response, there is some direct empirical evidence that GP training and support for use of alcohol screening and brief interventions may not be effective at changing clinical practice (Aalto et al., 2003; Andreasson, Hjalmarsson, & Rehnman, 2000; Beich et al., 2002; Kaner, Lock, McAvoy, Heather, & Gilvarry, 1999). Moreover, despite the inclusion of substance use teaching in undergraduate and vocational medical education over the past two decades, there is little evidence that GPs are identifying more at-risk drinkers (Section 1.3).

1.5 GP beliefs, attitudes, and experiences

As has been established, the preventive and health promotion approach to managing alcohol problems in primary care is endorsed by governmental and peak GP and public health organisations. The tools for this process appear effective in research settings. However, GPs consistently do not use these tools. What is the explanation for this phenomenon?

There is no accepted coherent answer that is apparent in the literature. A few researchers have explicitly questioned the rationale and feasibility of universal alcohol screening in general practice (Beich et al., 2003, 2007), but this is a minority opinion. The aforementioned predominant response in the literature has, and continues to, focus on GP-related factors (knowledge, skills and confidence). Categorisation of the barriers to implementation is typically limited mainly to clinician factors, some organisational factors, and occasional patient factors (Johnson, Jackson, Guillaume, Meier, & Goyder, 2011).

When a broader historical perspective is taken, the current problem appears to be a continuation of the lack of alignment between the enthusiasm for early detection and intervention of at-risk drinking by alcohol researchers, and clinical practice by GPs. This is an issue that has been recognised before (Murphy, 1980; Reid et al., 1986; Thom & Téllez, 1986; P. Wallace & Haines, 1985). When framed in this manner, it is clear that the perceptions of both alcohol researchers and clinical GPs need to be considered. Interestingly, prior to the study by Thom and Téllez (1986), the views of GPs in this field were not only absent in the literature, they were not sought. The authors opined:

> *"It has often been assumed that the failure of general practitioners to enter whole heartedly into involvement with these problems is evidence simply of ignorance, prejudice or wrong thinking. More often than not, however, discussion is based on the open or latent assumption that the general practitioner's active involvement with this type of patient is 'a good thing'." (Ibid, p. 405)*

When reviewed, the perceptions of GPs on the use of alcohol screening questions provide some illumination. A noteworthy feature of the published literature on alcohol screening questionnaires from a GP perspective is the near complete absence of evidence of GP acceptability of these tools. This is remarkable given that these tools were designed for GPs to use. There appears to be only one contemporary report of positive GP impressions of using an alcohol screening tool, and this was embedded within a general lifestyle questionnaire, rather than as a stand-alone instrument (Goodyear-Smith et al., 2004).

The remainder of the published literature on GP opinions on the use of alcohol screening questionnaires is negative in tone. Representative of these opinions are GP experiences of using the AUDIT in the study by Beich et al. (2002) – it impaired the patient-centred approach, was awkward to implement, and disturbed the usual collaboration between patient and doctor. These GP beliefs, attitudes and experiences on alcohol screening questionnaires appear consistently across a number of other studies. Other common perceptions and experiences were that screening with a questionnaire was seen as intrusive by some patients, the use of these tools were overly time and resource intensive, and that universal/routine screening of patients was impossible in pragmatic practice (Brady, Sibthorpe, Bailie, Ball, & Sumnerdodd, 2002; Johansson, Bendtsen, & Åkerlind, 2005; Nygaard & Aasland, 2011).

Crucially, GPs who have been involved in implementing alcohol screening (and brief interventions) in their practices were often the most critical. In Beich et al. (2002), "the general practitioners who volunteered in our study to implement a screening and brief intervention programme in their own practice could not subsequently recommend it" (p. 4). Similarly the GPs in Andreasson et al. (2000) perceived that the AUDIT was of little or no value to their practice, even after education and support.

An importantly observation is that many of the barriers that emerge in the research of GP perceptions seem to have been about issues that go beyond that of implementing a specific tool or program. Rather, they were related to inquiring about patient alcohol use within the consultation generally. The early work by Thom and Téllez (1986) of British GPs found that: (i) GPs were uncomfortable with intruding into what they perceived as a "normal everyday activity and part of the patient's 'private life'" (p. 407); (ii) GPs believed that alcohol diagnoses were sensitive due to the normality and desirability of drinking, and the stigma attached to problem drinking; (iii) GPs perceived that the detection and management of alcohol problems was a process of negotiation inside a long-term patient-doctor relationship – changing patient beliefs might take years; (iv) GPs believed that the difficulties they faced arose from societal beliefs and attitudes on drinking; and thus (v) they were sceptical of the value of additional education, "these approaches merely emphasized the role of medicine in responding to drinking problems while ignoring the importance of wider social influences on the direction and outcome of medical interventions" (p. 414); and moreover (vi) felt pessimistic that they could help patients who were detected – they believed that important factors to achieving success lay outside of their control.

There have been a number of subsequent studies. The issues of lack of time, fear of spoiling the patient-doctor relationship, maintenance of the patient's integrity, the sensitive nature of alcohol diagnoses, the context of the presenting agenda, pessimism of the effectiveness of interventions, the social dimension of drinking, stigma, shame, and untruthfulness of patient answers have been reported by GPs (Aira, Kauhanen, Larivaara, & Rautio, 2003; Arborelius & Damstrom Thakker, 1995; Beich et al., 2002; Johansson et al., 2005; Nygaard & Aasland, 2011).

1.6 Conclusion

GPs have a principle role in reducing alcohol-related harm, but the theoretical potential of identifying at-risk drinkers in primary care and delivering early interventions has not been met. Alcohol researchers in the past three decades have undertaken a major initiative to develop and validate alcohol screening tools and brief interventions. This has been successful insofar that these tools now exist and appear to be efficacious. However, this research does not appear to have resulted in substantial changes in actual clinical practice.

The perspectives of GPs are potentially informative, but have been largely overshadowed in the published literature. Their perceptions and experiences appear to have had limited influence in shaping the thinking in this field, and more pragmatically, on primary care practice guidelines.

The qualitative literature that explores GP beliefs and attitudes is relatively limited with the majority of the research based on British and Scandinavian GPs. This research suggests that sociological factors may be an important barrier to detection (Aira et al., 2003; Nygaard & Aasland, 2011; Thom & Téllez, 1986). It is possible that differences in Australian sociocultural beliefs about alcohol consumption will be reflected in the perceptions of detecting at-risk drinking in local GPs. Thus, this research project was focussed on three aims:

- Describing Australian GP perceptions of the detection and screening of at-risk drinking;

- Understanding the low uptake of alcohol screening questionnaires, and in particular, the AUDIT-C which is the tool suggested for use in Australia (Demirkol et al., 2011; Harris et al., 2009b)

- Understanding the low overall detection of at-risk drinking.

Part I – Literature Review

Appendices

Part I – Literature Review

2.1 Comparison of the NHMRC 2001 vs. 2009 guidelines

The following table describes the prevalence of at-risk drinking in the 2007 National Drug Strategy House Survey. The definition of at-risk drinking using the NHMRC 2001 and 2009 guideline definitions are compared (AIHW, 2008, Table 5.2, p. 32; 2011, Table 4.5, pp. 55-57).

Table 2 At-risk drinking in the 2007 NDSHS data

NHMRC 2001 guidelines		NHMRC 2009 guidelines	
Increased short-term risk (*monthly prevalence*)	20.4%	Increased single occasion risk (*at least monthly*) [†]	28.7%
		Increased single occasion risk (*at least weekly*) [†]	16.1%
Increased long-term risk (*monthly*)	10.3%	Increased lifetime risk	20.3%
Increased short- *OR* long-term risk (*monthly*)	22.1%	Increased single occasion *OR* lifetime risk [‡]	22.7% (see comment)

[†] The re-analysis of the 2007 NDSHS data in the 2010 NDSHS was not published in a manner that makes this comparison exactly analogous.

[‡] This is an estimate calculated using the following assumptions: (i) increased single occasion risk refers to "at least weekly", and (ii) that the proportion of individuals who drink "at least weekly" at increased single occasion risk, compared to all individuals who drink at increased single occasion risk, was the same between 2007 and 2010. This estimate can be directly compared to the "single occasion *OR* lifetime risk" in the 2010 NDSHS in Section 1.2.1.

Part I – Literature Review

2.2 List of alcohol screening questionnaires

Table 3 Alcohol screening questionnaires

Tool	Details	Year described
CAGE	The earliest brief screening questionnaire: 4 items (Ewing, 1984)	1970
MAST	"Michigan Alcoholism Screening Test": 25 items – structured interview and self-administered questionnaire (Selzer, 1971)	1971
SMAST	"Short" MAST: 13 questions (Selzer et al., 1975)	1975
AUDIT	"Alcohol Use Disorders Identification Test": 10 items; developed by the World Health Organisation (J. B. Saunders et al., 1993b)	1989
T-ACE	Brief questionnaire for at-risk drinking in pregnant women that includes 3 questions from CAGE: 4 items (Sokol, Martier, & Ager, 1989)	1989
TWEAK	Brief questionnaire for at-risk drinking in pregnant women that includes 3 questions from T-ACE, and 2 from MAST: 5 items (Chan, Pristach, Welte, & Russell, 1993; Russell et al., 1994)	1993/4
Five-Shot	Questionnaire comprised of 2 questions from the AUDIT, and 3 questions from CAGE: 5 items (Seppa, Lepisto, & Sillanaukee, 1998)	1998
AUDIT-C	The consumption items of the AUDIT: 3 items (Bush et al., 1998)	1998
AUDIT variants	AUDIT-PC: 5 items, 1997 (Piccinelli et al., 1997); AUDIT-3: 1 item, 1998 (Bush et al., 1998); AUDIT-4: 4 items, 2002 (Gual, Segura, Contel, Heather, & Colom, 2002)	
FAST	Based on 4 items from the AUDIT (Hodgson, Alwyn, John, Thom, & Smith, 2002)	2002
CHAT	"Case-finding and Help Assessment Tool": this lifestyle and mental health disorders screening tool includes two items on alcohol (Goodyear-Smith et al., 2004, 2008)	2004

Part I – Literature Review

References

Aalto, M., Pekuri, P., & Seppa, K. (2002). Primary health care professionals' activity in intervening in patients' alcohol drinking: a patient perspective. *Drug and Alcohol Dependence, 66*(1), 39-43. doi: 10.1016/S0376-8716(01)00179-X

Aalto, M., Pekuri, P., & Seppa, K. (2003). Primary health care professionals' activity in intervening in patients' alcohol drinking during a 3-year brief intervention implementation project. *Drug and Alcohol Dependence, 69*(1), 9-14. doi: 10.1016/S0376-8716(02)00228-4

Aertgeerts, B., Buntinx, F., Ansoms, S., & Fevery, J. (2001). Screening properties of questionnaires and laboratory tests for the detection of alcohol abuse or dependence in a general practice population. *British Journal of General Practice, 51*(464), 206-217.

Aira, M., Kauhanen, J., Larivaara, P., & Rautio, P. (2003). Factors influencing inquiry about patients' alcohol consumption by primary health care physicians: qualitative semi-structured interview study. *Family Practice, 20*(3), 270-275. doi: 10.1093/fampra/cmg307

Andreasson, S., Hjalmarsson, K., & Rehnman, C. (2000). Implementation and dissemination of methods for prevention of alcohol problems in primary health care: a feasibility study. *Alcohol and Alcoholism, 35*(5), 525-530. doi: 10.1093/alcalc/35.5.525

Arborelius, E., & Damstrom Thakker, K. (1995). Why is it so difficult for general practitioners to discuss alcohol with patients? *Family Practice, 12*(4), 419-422.

Australian Institute of Health and Welfare. (2008). *2007 National Drug Strategy Household Survey: detailed findings. Drug statistics series no. 22. Cat. no. PHE 107.* Canberra: Australian Institute of Health and Welfare.

Australian Institute of Health and Welfare. (2011). *2010 National Drug Strategy Household Survey report. Drug statistics series no. 25. Cat. no. PHE 145.* Canberra: Australian Institute of Health and Welfare.

Begg, S., Vos, T., Barker, B., Stevenson, C., & Lopez, A. D. (2007). *The burden of disease and injury in Australia 2003. PHE 82.* Canberra: Australian Institute of Health and Welfare.

Beich, A., Gannik, D., & Malterud, K. (2002). Screening and brief intervention for excessive alcohol use: qualitative interview study of the experiences of general practitioners. *British Medical Journal, 325*(7369), 870. doi: 10.1136/bmj.325.7369.870

Beich, A., Gannik, D., Saelan, H., & Thorsen, T. (2007). Screening and brief intervention targeting risky drinkers in Danish general practice – a pragmatic controlled trial. *Alcohol and Alcoholism, 42*(6), 593-603. doi: 10.1093/alcalc/agm063

Beich, A., Thorsen, T., & Rollnick, S. (2003). Screening in brief intervention trials targeting excessive drinkers in general practice: systematic review and meta-analysis. *British Medical Journal, 327*(7414), 536-542. doi: 10.1136/bmj.327.7414.536

Bernadt, M. W., Mumford, J., Taylor, C., Smith, B., & Murray, R. M. (1982). Comparison of questionnaire and laboratory tests in the detection of excessive drinking and alcoholism. *Lancet, 1*(8267), 325-328. doi: 10.1016/S0140-6736(82)91579-3

Berner, M. M., Harter, M., Kriston, L., Lohmann, M., Ruf, D., Lorenz, G., & Mundle, G. (2007a). Detection and management of alcohol use disorders in German primary care influenced by non-clinical factors. *Alcohol and Alcoholism, 42*(4), 308-316. doi: 10.1093/alcalc/agm013

Berner, M. M., Kriston, L., Bentele, M., & Harter, M. (2007b). The alcohol use disorders identification test for detecting at-risk drinking: A systematic review and meta-analysis. *Journal of Studies on Alcohol and Drugs, 68*(3), 461-473.

Bradley, K. A., DeBenedetti, A. F., Volk, R. J., Williams, E. C., Frank, D., & Kivlahan, D. R. (2007). AUDIT-C as a brief screen for alcohol misuse in primary care. *Alcoholism, Clinical and Experimental Research, 31*(7), 1208-1217. doi: 10.1111/j.1530-0277.2007.00403.x

Bradley, K. A., Kivlahan, D. R., & Williams, E. C. (2009). Brief approaches to alcohol screening: practical alternatives for primary care. *Journal of General Internal Medicine, 24*(7), 881-883. doi: 10.1007/s11606-009-1014-9

Brady, M., Sibthorpe, B., Bailie, R., Ball, S., & Sumnerdodd, P. (2002). The feasibility and acceptability of introducing brief intervention for alcohol misuse in an urban aboriginal medical service. *Drug and Alcohol Review, 21*(4), 375-380. doi: 10.1080/0959523021000023243

Britt, H., Miller, G. C., Charles, J., Henderson, J., Bayram, C., Pan, Y., . . . Fahridin, S. (2010a). *General practice activity in Australia 2009-10. General practice series no. 27. Cat. no. GEP 27.* Canberra: Australian Institute of Health and Welfare.

Britt, H., Miller, G. C., Charles, J., Henderson, J., Bayram, C., Valenti, L.,
. . . Chambers, T. (2010b). *General practice activity in Australia
2000-01 to 2009-10: 10 year data tables. General practice
series no. 28. Cat. no. GEP 28*. Canberra: Australian Institute of
Health and Welfare.

Buchsbaum, D. G., Buchanan, R. G., Poses, R. M., Schnoll, S. H., &
Lawton, M. J. (1992). Physician detection of drinking problems
in patients attending a general medicine practice. *Journal of
General Internal Medicine, 7*(5), 517-521.

Bush, K. R., Babor, T. F., Kivlahan, D. R., McDonell, M. B., Fihn, S. D., &
Bradley, K. A. (1998). The AUDIT alcohol consumption
questions (AUDIT-C): an effective brief screening test for
problem drinking. Ambulatory Care Quality Improvement
Project (ACQUIP). Alcohol Use Disorders Identification Test.
Archives of Internal Medicine, 158(16), 1789-1795. doi: 10-
1001/pubs.Arch Intern Med.-ISSN-0003-9926-158-16-ioi70602

Centers for Disease Control and Prevention. (2012). Alcohol and Public
Health - Frequently Asked Questions.   Retrieved October 8,
2012, from http://www.cdc.gov/alcohol/faqs.htm

Chan, A. W., Pristach, E. A., Welte, J. W., & Russell, M. (1993). Use of
the TWEAK test in screening for alcoholism/heavy drinking in
three populations. *Alcoholism, Clinical and Experimental
Research, 17*(6), 1188-1192. doi: 10.1111/j.1530-
0277.1993.tb05226.x

Chikritzhs, T. N., Allsop, S. J., Moodie, A. R., & Hall, W. D. (2010). Per
capita alcohol consumption in Australia: will the real trend
please step forward? *Medical Journal of Australia, 193*(10),
594-597.

D'Amico, E. J., Paddock, S. M., Burnam, A., & Kung, F. Y. (2005). Identification of and guidance for problem drinking by general medical providers: results from a national survey. *Medical Care, 43*(3), 229-236.

Deehan, A., Templeton, L., Taylor, C., Drummond, C., & Strang, J. (1998). Low detection rates, negative attitudes and the failure to meet the "Health of the Nation" alcohol targets: findings from a national survey of GPs in England and Wales. *Drug and Alcohol Review, 17*(3), 249-258. doi: 10.1080/09595239800187081

Demirkol, A., Haber, P., & Conigrave, K. (2011). Problem drinking - detection and assessment in general practice. *Australian Family Physician, 40*(8), 570-574.

Department of Health and Ageing. (2009). The Australian Standard Drink. Retrieved October 8, 2012, from http://www.alcohol.gov.au/internet/alcohol/publishing.nsf/Content/standard

Durand, M. A. (1994). General practice involvement in the management of alcohol misuse: dynamics and resistances. *Drug and Alcohol Dependence, 35*(3), 181-189.

Engdahl, B., & Nilsen, P. (2011). Receiving an alcohol enquiry from a physician in routine health care in Sweden: a population-based study of gender differences and predictors. *Int J Environ Res Public Health, 8*(5), 1296-1307. doi: 10.3390/ijerph8051296

Ewing, J. A. (1984). Detecting alcoholism. The CAGE questionnaire. *JAMA, 252*(14), 1905-1907. doi: 10.1001/jama.1984.03350140051025

Family Medicine Research Centre. (2012). Bettering the Evaluation and Care of Health (BEACH). Retrieved 27 November, 2012, from http://sydney.edu.au/medicine/fmrc/beach/

Friedmann, P. D., McCullough, D., Chin, M. H., & Saitz, R. (2000). Screening and intervention for alcohol problems. A national survey of primary care physicians and psychiatrists. *Journal of General Internal Medicine, 15*(2), 84-91. doi: 10.1046/j.1525-1497.2000.03379.x

Goodyear-Smith, F., Arroll, B., Sullivan, S., Elley, R., Docherty, B., & Janes, R. (2004). Lifestyle screening: development of an acceptable multi-item general practice tool. *New Zealand Medical Journal, 117*(1205), U1146.

Goodyear-Smith, F., Coupe, N. M., Arroll, B., Elley, C. R., Sullivan, S., & McGill, A. T. (2008). Case finding of lifestyle and mental health disorders in primary care: validation of the 'CHAT' tool. *British Journal of General Practice, 58*(546), 26-31. doi: 10.3399/bjgp08X263785

Gual, A., Segura, L., Contel, M., Heather, N., & Colom, J. (2002). Audit-3 and audit-4: effectiveness of two short forms of the alcohol use disorders identification test. *Alcohol and Alcoholism, 37*(6), 591-596. doi: 10.1093/alcalc/37.6.591

Harris, M., Bennett, J., Del Mar, C., Fasher, M., Foreman, L., Furler, J., . . . Snowdon, T. (2009a). AUDIT-C *Guidelines for preventive activities in general practice* (7th ed., pp. 86-87). South Melbourne: The Royal Australian College of General Practitioners.

Harris, M., Bennett, J., Del Mar, C., Fasher, M., Foreman, L., Furler, J., . . . Snowdon, T. (2009b). Early detection of problem drinking *Guidelines for preventive activities in general practice* (7th ed., pp. 37-39). South Melbourne: The Royal Australian College of General Practitioners.

Hodgson, R., Alwyn, T., John, B., Thom, B., & Smith, A. (2002). The FAST Alcohol Screening Test. *Alcohol and Alcoholism, 37*(1), 61-66. doi: 10.1093/alcalc/37.1.61

Johansson, K., Bendtsen, P., & Åkerlind, I. (2005). Factors influencing GPs' decisions regarding screening for high alcohol consumption: a focus group study in Swedish primary care. *Public Health, 119*(9), 781-788. doi: 10.1016/j.puhe.2004.12.006

Johnson, M., Jackson, R., Guillaume, L., Meier, P., & Goyder, E. (2011). Barriers and facilitators to implementing screening and brief intervention for alcohol misuse: a systematic review of qualitative evidence. *Journal of Public Health, 33*(3), 412-421. doi: 10.1093/pubmed/fdq095

Kaner, E. F., Dickinson, H. O., Beyer, F., Pienaar, E., Schlesinger, C., Campbell, F., . . . Heather, N. (2009). The effectiveness of brief alcohol interventions in primary care settings: a systematic review. *Drug and Alcohol Review, 28*(3), 301-323. doi: 10.1111/j.1465-3362.2009.00071.x

Kaner, E. F., Lock, C. A., McAvoy, B. R., Heather, N., & Gilvarry, E. (1999). A RCT of three training and support strategies to encourage implementation of screening and brief alcohol intervention by general practitioners. *British Journal of General Practice, 49*(446), 699-703.

Leckman, A. L., Umland, B. E., & Blay, M. (1984). Prevalence of alcoholism in a family practice center. *Journal of Family Practice, 18*(6), 867-870.

Meneses-Gaya, C. d., Crippa, J. A., Zuardi, A. W., Loureiro, S. R., Hallak, J. E., Trzesniak, C., . . . Martin-Santos, R. (2010). The fast alcohol screening test (FAST) is as good as the AUDIT to screen alcohol use disorders. *Substance Use and Misuse, 45*(10), 1542-1557. doi: 10.3109/10826081003682206

Meneses-Gaya, C. d., Zuardi, A. W., Loureiro, S. R., & Crippa, J. A. S. (2009). Alcohol Use Disorders Identification Test (AUDIT): an updated systematic review of psychometric properties. *Psychology & Neuroscience, 2*, 83-97. doi: 10.3922/j.psns.2009.1.12

Ministerial Council on Drug Strategy. (2006). *National Alcohol Strategy 2006-2011*. Canberra: Commonwealth of Australia.

Murphy, H. B. (1980). Hidden barriers to the diagnosis and treatment of alcoholism and other alcohol misuse. *Journal of Studies on Alcohol, 41*(5), 417-428.

National Health and Medical Research Council. (2001). *Australian Alcohol Guidelines Health Risks and Benefits*. Canberra: Commonwealth of Australia.

National Health and Medical Research Council. (2009). *Australian Guidelines to Reduce Health Risks from Drinking Alcohol*. Canberra: Commonwealth of Australia.

National Institute for Health and Clinical Excellence. (2011). *Alcohol-use disorders: diagnosis, assessment and management of harmful drinking and alcohol dependence*. London: National Institute for Health and Clinical Excellence.

Nygaard, P., & Aasland, O. G. (2011). Barriers to implementing screening and brief interventions in general practice: findings from a qualitative study in Norway. *Alcohol and Alcoholism, 46*(1), 52-60. doi: 10.1093/alcalc/agq073

Nygaard, P., Paschall, M. J., Aasland, O. G., & Lund, K. E. (2010). Use and Barriers to Use of Screening and Brief Interventions for Alcohol Problems Among Norwegian General Practitioners. *Alcohol and Alcoholism, 45*(2), 207-212. doi: 10.1093/alcalc/agq002

Piccinelli, M., Tessari, E., Bortolomasi, M., Piasere, O., Semenzin, M., Garzotto, N., & Tansella, M. (1997). Efficacy of the alcohol use disorders identification test as a screening tool for hazardous alcohol intake and related disorders in primary care: A validity study. *British Medical Journal (Clinical Research Ed.), 314*(7078), 420-424.

Proude, E. M., Britt, H., Valenti, L., & Conigrave, K. M. (2006). The relationship between self-reported alcohol intake and the morbidities managed by GPs in Australia. *BMC Family Practice, 7*, 17. doi: 10.1186/1471-2296-7-17

Reid, A. L., Webb, G. R., Hennrikus, D., Fahey, P. P., & Sanson-Fisher, R. W. (1986). Detection of patients with high alcohol intake by general practitioners. *British Medical Journal (Clinical Research Ed.), 293*(6549), 735-737.

Reinert, D. F., & Allen, J. P. (2007). The Alcohol Use Disorders Identification Test: An update of research findings. *Alcoholism, Clinical and Experimental Research, 31*(2), 185-199. doi: 10.1111/j.1530-0277.2006.00295.x

Richmond, R., Kehoe, L., Heather, N., Wodak, A., & Webster, I. (1996). General practitioners' promotion of healthy life styles: what patients think. *Australian and New Zealand Journal of Public Health, 20*(2), 195-200.

The Royal Australian College of General Practitioners. (2004). *Smoking, Nutrition, Alcohol and Physical activity (SNAP), A population health guide to behavioural risk factors in general practice*. South Melbourne: The Royal Australian College of General Practitioners.

Russell, M., Martier, S. S., Sokol, R. J., Mudar, P., Bottoms, S., Jacobson, S., & Jacobson, J. (1994). Screening for pregnancy risk-drinking. *Alcoholism, Clinical and Experimental Research, 18*(5), 1156-1161. doi: 10.1111/j.1530-0277.1994.tb00097.x

Rydon, P., Redman, S., Sanson-Fisher, R. W., & Reid, A. L. (1992). Detection of alcohol-related problems in general practice. *Journal of Studies on Alcohol, 53*(3), 197-202.

Sanders, J. L. (2011). A distinct language and a historic pendulum: the evolution of the Diagnostic and Statistical Manual of Mental Disorders. *Archives of Psychiatric Nursing, 25*(6), 394-403. doi: 10.1016/j.apnu.2010.10.002

Saunders, J. B., Aasland, O. G., Amundsen, A., & Grant, M. (1993a). Alcohol Consumption and Related Problems among Primary Health Care Patients: WHO Collaborative Project on Early Detection of Persons with Harmful Alcohol Consumption – I. *Addiction, 88*(3), 349-362. doi: 10.1111/j.1360-0443.1993.tb00822.x

Saunders, J. B., Aasland, O. G., Babor, T. F., Delafuente, J. R., & Grant, M. (1993b). Development of the Alcohol Use Disorders Identification Test (AUDIT): WHO Collaborative Project on Early Detection of Persons with Harmful Alcohol Consumption – II. *Addiction, 88*(6), 791-804. doi: 10.1111/j.1360-0443.1993.tb02093.x

Saunders, W. M., & Kershaw, P. W. (1980). Screening tests for alcoholism – findings from a community study. *British Journal of Addiction, 75*(1), 37-41.

Science and Technology Committee. (2012). *Alcohol guidelines - Eleventh Report of Session 2010-12*. London: The Stationary Office Limited.

Seale, J. P., Boltri, J. M., Shellenberger, S., Velasquez, M. M., Cornelius, M., Guyinn, M., . . . Sumner, H. (2006). Primary care validation of a single screening question for drinkers. *Journal of Studies on Alcohol, 67*(5), 778-784.

Selzer, M. L. (1971). The Michigan alcoholism screening test: the quest for a new diagnostic instrument. *American Journal of Psychiatry, 127*(12), 1653-1658.

Selzer, M. L., Vinokur, A., & van Rooijen, L. (1975). A self-administered Short Michigan Alcoholism Screening Test (SMAST). *Journal of Studies on Alcohol, 36*(1), 117-126.

Seppa, K., Lepisto, J., & Sillanaukee, P. (1998). Five-shot questionnaire on heavy drinking. *Alcoholism, Clinical and Experimental Research, 22*(8), 1788-1791. doi: 10.1111/j.1530-0277.1998.tb03981.x

Sokol, R. J., Martier, S. S., & Ager, J. W. (1989). The T-ACE questions: practical prenatal detection of risk-drinking. *American Journal of Obstetrics and Gynecology, 160*(4), 863-868; discussion 868-870.

Solberg, L. I., Maciosek, M. V., & Edwards, N. M. (2008). Primary care intervention to reduce alcohol misuse ranking its health impact and cost effectiveness. *American Journal of Preventive Medicine, 34*(2), 143-152. doi: 10.1016/j.amepre.2007.09.035

Spandorfer, J. M., Israel, Y., & Turner, B. J. (1999). Primary care physicians' views on screening and management of alcohol abuse: inconsistencies with national guidelines. *Journal of Family Practice, 48*(11), 899-902.

Thom, B., & Téllez, C. (1986). A difficult business: detecting and managing alcohol problems in general practice. *British Journal of Addiction, 81*(3), 405-418.

U.S. Preventive Services Task Force. (2004). Screening and behavioral counseling interventions in primary care to reduce alcohol misuse: recommendation statement.   Retrieved October 31, 2012, from http://www.uspreventiveservicestaskforce.org/uspstf/uspsdrin.htm

Vande Creek, L., Zachrich, R. L., & Scherger, W. E. (1982). The use of standardized alcoholism screening tests in family practice. *Family Practice Research Journal, 2*, 11-17.

Wallace, P., & Haines, A. (1985). Use of a Questionnaire in General Practice to Increase the Recognition of Patients with Excessive Alcohol Consumption. *British Medical Journal (Clinical Research Ed.), 290*(6486), 1949-1953.

Wallace, P. G., Brennan, P. J., & Haines, A. P. (1987). Are general practitioners doing enough to promote healthy lifestyle? Findings of the Medical Research Council's general practice research framework study on lifestyle and health. *British Medical Journal (Clinical Research Ed.), 294*(6577), 940-942.

Wallace, P. G., & Haines, A. P. (1984). General practitioner and health promotion: what patients think. *British Medical Journal (Clinical Research Ed.), 289*(6444), 534-536.

Yamada, K., Maeno, T., Waza, K., & Sato, T. (2008). Under-diagnosis of alcohol-related problems and depression in a family practice in Japan. *Asia Pacific Family Medicine, 7*(1), 3. doi: 10.1186/1447-056X-7-3

Part II

Journal Article

Part II – Journal Article

4.1 Journal guidelines for *Drug and Alcohol Review*

Drug and Alcohol Review is the journal of the Australasian Professional Society on Alcohol and other Drugs. The following manuscript guidelines are adapted from the "Author Guidelines" page on this journal's website. Only the details and requirements relevant to the submitted article are included.

Manuscript category – original papers
Reports of new research findings that make a significant contribution to knowledge (3000 word limit).

Requirements of submitted articles

Cover Letter
- The International Society of Addiction Journal Editors (ISAJE) Guidelines define authorship as substantial contribution to all aspects of the research design, analysis and interpretation of data and contribution to the intellectual content of the article. All authors must acknowledge these guidelines and be willing to take public responsibility for the content of the article.
- Papers are accepted for publication in the Journal on the understanding that the content has not been published or submitted for publication elsewhere. This must be stated in the cover letter.
- Authors must declare any financial support or relationships that may pose conflict of interest.

- If tables or figures have been reproduced from another source, a letter from the copyright holder (usually the publisher), stating authorisation to reproduce the material, must be attached to the cover letter.

Ethical Considerations

- ISAJE Ethical Practice Guidelines provide guidance to authors 'regarding ethical and procedural issues that affect the integrity of scientific publishing'. We ask that authors read and observe these guidelines especially in regard to study design and ethical approval, consent, authorship, conflict of interests, plagiarism and redundant publication.
- Authors must state that the protocol for the research project has been approved by a suitably constituted Ethics Committee of the institution within which the work was undertaken and that it conforms to the provisions of the Declaration of Helsinki (as revised in Tokyo 2004).

Manuscript Parts

Manuscripts should be presented in the following order:

(a) Title page
(b) Abstract and key words
(c) Text
(d) Acknowledgements
(e) References
(f) Appendices
(g) Figures

(h) Tables

(i) An ethics statement must be included in the Methods section.

Title page

The title page should contain:

1. The title of the paper – the title should be concise and informative, with no abbreviations
2. A short running title of no more than 40 characters
3. The full names, academic qualifications (e.g. BA, MSc, PhD etc.) and job position titles of the authors
4. The name of department(s) and institution(s) at which the work was carried out.
5. The full postal and email address, plus facsimile and telephone numbers, of the corresponding author.

Abstract and key words

The second page should carry a structured abstract of not more than 250 words, using the following headings: Introduction and Aims, Design and Methods, Results, Discussion and Conclusions

For the purposes of indexing, five key words, should be supplied below the abstract and should be taken from those recommended by the US National Library of Medicine's Medical Subject Headings (MeSH) browser list.

Part II – Journal Article

Text

The text of Original Papers should conform to the conventional structure for biomedical communications - introduction, methods, results and discussion.

We suggest authors follow guidelines for the discussion section of their paper, as reported in the British Medical Journal (abstract): (Docherty BMJ 1999)

- Statement of principal findings
- Strengths and weaknesses of the study
- Strengths and weaknesses in relation to other studies, discussing particularly any differences in results
- Meaning of the study: possible mechanisms and implications for clinicians or policymakers
- Unanswered questions and future research

Acknowledgements

All sources of support in the form of financial grants, equipment or drugs should be stated in the acknowledgement section. Contributions of colleagues or institutions can be acknowledged but personal thanks or appreciation of anonymous reviewers is not appropriate.

References

The Vancouver system of referencing should be used. References should be numbered consecutively in the order in which they are first mentioned in the text. Indicate in the text with Arabic numbers inside square brackets. In the reference list, cite the names of all authors. All citations mentioned in the text, tables or figures must be listed in the reference list.

Tables

Tables should be self-contained and complement, not duplicate, information contained in the text. Number tables consecutively in the text in Arabic numerals. Explanatory matter, including definition of abbreviations, should be placed in footnotes. Type tables on a separate page with the legend above; legends should be concise but comprehensive - the table, legend and footnotes must be understandable without reference to the text. Vertical lines should not be used to separate columns. Column headings should be brief, with units of measurement in parentheses.

Manuscript Style

We encourage high quality writing and refer our authors to one of the many writing guides available online such as Kipling's Guide to Writing a Scientific Paper. Manuscripts should follow the style of the Vancouver agreement detailed in the International Committee of Medical Journal Editors' revised Uniform Requirements for Manuscripts Submitted to Biomedical Journals: Writing and Editing for Biomedical Publication.

Manuscripts are submitted online in Word, double spaced, page number in top right-hand corner, do not use 'Enter' at the end of lines within a paragraph.

- Spelling: The Journal uses English (UK) spelling.

- Units: All measurements must be given in SI or SI-derived units with traditional units in parentheses.
- Abbreviations: An explanation of all abbreviations should be given on the first occasion of their use, with the exception of abbreviations of certain standard units of measurement and statistical measures of variation such as SD and SEM which may be used without explanation.

Part II – Journal Article

Manuscript

Part II – Journal Article

5.1 Cover letter

Chun Wah Michael Tam
School of Public Health and Community Medicine
The University of New South Wales
UNSW Sydney, NSW 2052, Australia

SCHOOL OF PUBLIC HEALTH
AND COMMUNITY MEDICINE

Drug and Alcohol Review
John Wiley & Sons Singapore Pte. Ltd.
1 Fusionopolis Walk
#07-01 Solaris South Tower
Singapore 138 628

8 December 2012

Dear Editor,

"Australian general practitioner perceptions of the detection and screening of at-risk drinking, and the role of the AUDIT-C"

At-risk drinking is common in Australia and general practitioners (GPs) do not appear to use screening questionnaires like the AUDIT-C despite recommendations. In this study, we used qualitative methods to describe GP beliefs, and to seek an explanation for this phenomenon. We uncovered the major influence of sociocultural factors on the barriers to detection of at-risk drinking in primary care and discuss the implications of these findings. Specifically, recommendations suggesting GP use of these tools are unlikely to be successful at improving detection of at-risk drinking in the current Australian context.

The content of this paper has not been published or submitted for publication elsewhere.

Contact details:
Corresponding author: Dr C W Michael Tam, BSc(Med) MBBS FRACGP
E-mail: m.tam@unsw.edu.au (preferred)
Mob: +61 412 704 158
Phone: +61 2 9385 2520
Fax: +61 2 9313 6185

Co-authors: Prof Nicholas Zwar, MBBS MPH PhD FRACGP
E-mail: n.zwar@unsw.edu.au

Dr Roslyn Markham, PhD
E-mail: ros.markham@nswiop.nsw.edu.au

Thank you for your kind attention and considering our article for publication in *Drug and Alcohol Review*.

Yours sincerely,

Michael Tam
Lecturer in Primary Care and General Practitioner

School of Public Health and Community Medicine, UNSW Sydney 2052, Australia
Mob: +61 412 704 158 | Ph: +61 2 9385 2520 | Fax: +61 2 9313 6185 | e-mail: m.tam@unsw.edu.au

Part II – Journal Article

5.2 Title page

Full title: *Australian general practitioner perceptions of the detection and screening of at-risk drinking, and the role of the AUDIT-C*

Short title: Australian GP perceptions of at-risk drinking

Corresponding author

Dr Chun Wah Michael Tam [a,b]

Qualification:	BSc(Med) MBBS FRACGP
Position:	Lecturer in Primary Care
Contact:	Level 3 Samuels Building, School of Public Health and Community Medicine
	The University of New South Wales
	UNSW Sydney, NSW 2052, Australia
	E-mail: m.tam@unsw.edu.au (preferred)
	Mob: +61 412 704 158
	Phone: +61 2 9385 2520
	Fax: +61 2 9313 6185

Co-authors

Prof Nicholas Zwar [a]

Qualification:	MBBS MPH PhD FRACGP
Position:	Professor of General Practice
Contact:	n.zwar@unsw.edu.au

Dr Roslyn Markham [b]

Qualification:	PhD
Position:	Research Postgraduate Coordinator
Contact:	ros.markham@nswiop.nsw.edu.au

(a) School of Public Health and Community Medicine, The University of New South Wales, Sydney, NSW Australia

(b) The New South Wales Institute of Psychiatry, North Parramatta, NSW Australia

5.3 Abstract

Introduction and Aims: At-risk drinking is common in Australia. Validated screening tools such as the AUDIT-C have been promoted to general practitioners (GPs), but appear rarely used. Detection of at-risk drinking in primary care remains low. We sought to describe Australian GP perceptions of the detection and screening of at-risk drinking; to understand the low uptake of alcohol screening questionnaires, and in particular, the role of the AUDIT-C; and the overall low detection of at-risk drinking.

Design and Methods: Focus group interviews of four groups of GPs and GP trainees, with a total of 19 participants, were conducted in metropolitan Sydney between August and October 2011.   Audio recordings were transcribed and analysed using grounded theory methodology.

Results: Four major factors arose: (i) perceptions of the detection of at-risk drinking, (ii) sociocultural attitudes towards drinking, (iii) dynamics of patient-doctor interactions, and (iv) perceptions of alcohol screening and the AUDIT-C.

Discussion and Conclusion: Sociocultural factors appear to have a key influence on the major barriers to detection of at-risk drinking in primary care. These barriers were: community stigma and stereotypes of "problem" drinking, GP perceptions of unreliable patient alcohol use histories, and the perceived threat to the patient-doctor relationship from alcohol use assessment.   Alcohol screening questionnaires such as the AUDIT-C are not designed to address these factors and barriers. In the current context, it is unlikely that approaches that focus on the use of these tools will be effective at improving detection of at-risk drinking by GPs.

MeSH keywords

"Alcohol drinking", "Primary Health Care", "General Practitioners", "Substance Abuse Detection", "Health Knowledge, Attitudes, Practice"

Part II – Journal Article

5.4 Main text

Introduction

At-risk drinking is common in Australia. An estimated 3 in 10 adult patients presenting to general practitioners (GPs) drink at levels that place them at increased risk of alcohol-related harm [1]. Primary care has been identified as an important setting for the prevention and treatment of alcohol-related health issues [2]. Brief interventions delivered in this setting are effective in assisting patients reduce their alcohol use [3]. However, patients with at-risk drinking must be identified for brief interventions to be offered.

There is evidence dating back some years that Australian GPs did not identify the majority of these patients [4, 5]. This appears to be an enduring problem in contemporary general practice internationally [6-8]. Data from the Bettering the Evaluation and Care of Health (BEACH) Program suggests that alcohol counselling or advice is rarely given by Australian GPs – only in 4 in 1000 patient encounters [1, 9]. It is probable that the majority of at-risk drinkers continue not to be identified.

Alcohol screening questionnaires have been widely promoted as a means to address the problem of under-detection and the routine use of these tools is a feature of primary care treatment guidelines [10-13]. The Alcohol Use Disorders Identification Test (AUDIT) is the most studied and best validated screening questionnaire in primary care [14-16]. However, it has been found to be awkward to use, impaired the patient-centred approach, and imposed too high a workload in practice [17, 18]. Simpler questionnaires such as the AUDIT-C (based on the first three items of the AUDIT) have been proposed as effective validated alternatives [19, 20]. Australian guidelines encourage GPs to use this tool [10].

Despite their apparent effectiveness, there is debate on the acceptability of alcohol screening questionnaires to GPs. There are older [21] and more recent [22] reports of positive GP impressions of these tools. At the same time, other studies demonstrate that screening questionnaires are rarely used, even after extensive education and support has been provided [23-27].

There has been considerable qualitative exploration of the barriers to alcohol screening in primary care, but not in the Australian setting [28], which may differ from other cultural contexts. In this study we describe the beliefs and attitudes of Australian GPs and GP trainees on the detection and screening of at-risk drinking, with a focus on the AUDIT-C. We aimed to understand the reasons for the low GP uptake of alcohol screening questionnaires, and the overall low detection of at-risk drinking.

Methods

We used Straussian grounded theory, a qualitative research method that is exploratory, seeks to generate theory from analysis of data, and is suited to examining the underlying social processes of specific phenomena [29]. Its elements include: (i) coding of collected data into categories, (ii) an iterative process where earlier analyses inform further data collection, (iii) constant comparison in the analytic process to discover the conceptual interactions between categories, and (iv) the development of theory that emerges from and is grounded in the data.

Focus groups – the practices and participants

A convenience sample of GP teaching practices in Sydney affiliated with either the University of New South Wales, or GP Synergy Ltd. (a provider of GP vocational training) was approached to participate in the study.

Although the planned size of each focus group was 4-8 participants, a number of barriers were experienced in recruitment. Some practices had an insufficient number of interested participants. Finding a mutually convenient time of sufficient length (45 to 60 minutes) for the doctors to meet was difficult.

A total of nine group practices were approached and four agreed to participate. As theoretical saturation had been reached with this number of focus group interviews, no further practices were contacted. The focus groups were held between August and October 2011. They ranged from 2 to 7 participants each, with a total of 19 participants. Table 1 lists the practice and focus group characteristics.

All focus group interviews were moderated by the first author (Tam). A semi-structured approach was employed with five questions used as triggers if the issues did not arise naturally in discussion (Table 2). The participants were given a printed copy the AUDIT-C questionnaire [30] before the last trigger question. Each focus group was digitally recorded, transcribed, and de-identified. Participant demographics were collected by a short questionnaire. Table 3 summarises the characteristics of the participants.

Data analysis

Analysis of the transcripts, codes, categories and themes were performed using QSR International Nvivo 9 software. Themes emerging from earlier focus groups were introduced and explored in subsequent ones. Throughout the process the team met at regular intervals to discuss interpretations of the data though the coding was predominantly performed by the first author (Tam). He monitored coding consistency by continually reviewing and comparing the coding categories and sub-categories. The relationship between dominant themes and the development of theory was assisted with the use of diagrams and visualising the data with tree-maps within Nvivo 9.

Ethics approval

This study received approval from the University of NSW Medical and Community Human Research Ethics Advisory Panel, and the NSW Institute of Psychiatry Human Research Ethics Committee.

Results

Our analysis identified four major areas that help explain the low rates of detection of at-risk drinking in Australian primary care, and the lack of use of alcohol screening questionnaires. The sequence of the following themes and topics reflects the conversational flow of the group interviews. The GPs started with dialogue on their beliefs and attitudes on detecting at-risk drinking. They reflected on and debated the wider role of societal attitudes and sociocultural contexts, as well as focussing on the issues at the level of the consultation. Lastly, they specifically discussed their views on screening when they were presented with the AUDIT-C.

1. GP beliefs and attitudes on detection of at-risk drinking

Focus groups began with each participant asked to estimate the proportion of Australian adults presenting to GPs who were at-risk drinkers. The average of the estimates was 31%. Leading on from the initial discussion of prevalence, almost all GPs agreed that the detection of at-risk drinking was important:

> ...And I think it's certainly important ... as ... doctors taking care of the Australian population. That's [alcohol use] certainly ... one of the things we should be trying to identify and educate our patients about and hopefully be able to engage them in modifying their behaviour when they're willing... (D18)

Most GPs perceived that the accuracy of the assessment was crucial:

> ...But I also think people will say, "I only drink ... I don't drink very often," but that could be two or three big nights on the weekend which we think is excessive drinking. So I think it's really important to try ... to clarify for them exactly what they do... (D8)

There was a general view that alcohol use histories from patients were unreliable and typically under-estimated. This was perceived as a major barrier to detection of at-risk drinking:

> ...I think people are extraordinarily resourceful in the ways in which they're able to potentially cover up behaviour they don't want others to see. So, people are likely to under-report the amount that they're drinking, either knowingly or subconsciously... (D18)

Even when GPs identified a patient with at-risk drinking, they seemed reluctant to label it in clinical records. Some elaborated that at-risk drinking was not a diagnosis or disease, so shouldn't be labelled. Others were concerned with issues of patient confidentiality and the potential for the labels to be passed on to third parties, e.g., to insurers. Many GPs were concerned of being perceived as judgemental by patients. Nevertheless, there was a broad agreement that alcohol use is conceptually within the scope of health issues that can be addressed in the GP setting.

Some GPs expressed their sense of ineffectiveness in promoting behaviour change:

> ...I'm not that bad at detecting it. I'm bad at doing anything about it. As in ... I honestly don't know sometimes if I find out the answer whether it makes that much of a difference of what they end up, or what I end up doing... (D11)

2. Impact of social and cultural attitudes

Sociocultural attitudes and their impacts emerged strongly as a theme in all groups. It was widely perceived that some patterns of drinking that were risky were "normal" and even expected in many Australian community contexts:

> ...it's very socially acceptable to drink... It's completely normal for a teenager to come out of school at the age of 17 and start, you know, having binge drinks with his mates at bars and clubs... For girls to be given free drinks by bouncers and ... bartenders at nightclubs... (D16)

These GPs believed that at-risk drinking labels were perceived to be shameful by the community, and so individuals reject these labels for their personal drinking patterns:

> It's socially unacceptable to say you're a heavy drinker, but it's actually socially acceptable to be a heavy drinker. That's the problem... (D11)

In their discussions, the GPs described three main impacts on patient beliefs and behaviour arising from these sociocultural attitudes: (i) some patients were defensive about or hide their use of alcohol, (ii) others did not recognise the health risks from drinking, and (iii) drinking guidelines were not always perceived as "reasonable":

> ...Because then everybody who saw themselves as a perfectly reasonable person without an alcohol

> *problem, if you drank that much [referring to National Health and Medical Research Council guidelines]... they became a binge drinker, and I think that people... weren't interested to... they didn't listen any more. They just felt that was such an unreasonable label that they didn't want to listen to what it might be based on. (D8)*

3. Dynamics of patient-doctor interactions

Many GPs indicated that asking patients about their alcohol use was a potential threat to the patient-doctor relationship. At times, they were reluctant to introduce this topic into the consultation, being self-conscious about being perceived as moralising by the patient. The situation most frequently identified as challenging was a presenting problem that was seen as unrelated to alcohol use:

> *...And someone with a cold, unless they're a new patient ... it wouldn't be something [alcohol use] that you'd necessarily ask about... And they might be a bit affronted if you did, if you couldn't figure a way of bringing it in. (D6)*

However, assessment of alcohol use was perceived as rather less threatening to the relationship when the patient was new to the practice, when it was part of an explicit "health check", or when completing a computer health record:

> *...I've taken to the habit of updating the smoking and the alcohol history in the last year... and they're quite*

> *happy to be asked it, especially when you say, "Look,*
> *I'm just doing a refresh of your file [electronic record]*
> *and updating of all the information." (D12)*

Furthermore, there seemed to be presentations that actively triggered assessment of alcohol use. This was described to be the case for certain blood test abnormalities (elevated gamma-GT) and mental illness by many GPs:

> *...It often comes up in mental health consults, if they're*
> *depressed ... it's almost like an open door to ask a little*
> *about alcohol then. (D3)*

Several GPs identified other patient presentations that facilitated assessment, such as hypertension, traumatic accidents, and pregnancy and fertility.

4. GP beliefs and attitudes on alcohol screening and the AUDIT-C

Most GPs knew of the CAGE questionnaire [31], and a few were aware of the AUDIT and AUDIT-C. The universal experience, however, was that they either rarely or never used these tools. Those who had experience using these questionnaires did not use them for the purpose of routine screening. Rather, it had either been in a research setting (participating in the BEACH Program), or they used the questions as a framework for exploring alcohol use with patients they had already identified with an alcohol use disorder:

> *...Yeah, I suppose the times I've done it [CAGE] is just...*
> *actually getting the patient to engage with why I think*
> *they've got an alcohol problem. (D12)*

When the GPs were given the AUDIT-C questionnaire, the overwhelming and uniform perception was that it would over-identify patients with at-risk drinking. The questions were considered reasonable, but the scoring was met with marked scepticism.

> *...it's [detecting] a lot of people that probably are not*
> *really having a problem... it seems silly... (D17)*

Several GPs elaborated that although they believed that screening for at-risk drinking was good in theory, it needed to be performed in all patients to be useful. The pragmatic consensus was that routine screening with questionnaires was too difficult to implement, and that they could not and would not perform it consistently in practice.

Discussion

Our findings suggest that sociocultural factors have a key influence on the detection of at-risk drinking. The beliefs and attitudes of individual patients, doctors, and the expectations of medical consultations, sit within sociocultural contexts and are shaped by them [32]. There appears to be three linked major conceptual barriers to the detection of at-risk drinking by these GPs: (i) community stigma and stereotypes of "problem" drinking, (ii) GP perceptions of unreliable patient alcohol use histories, and (iii) the perceived threat to the patient-doctor relationship from alcohol use assessment.

The first major barrier, community stigma and stereotypes of "problem" drinking, has important direct effects on the detection of at-risk drinking. Consistent with our results, previous studies have also found that GPs were concerned about the possibility of being seen as "moralists" [33-35], and felt discussions on alcohol required particular sensitivity [36, 37]. This may explain GP reluctance to label patients who were at-risk drinkers.

The second major barrier, GP perceptions of the unreliability of patient alcohol use histories, can be conceptualised as a consequence of stigma. Individuals with alcohol dependence are particularly stigmatised compared to those with other mental health conditions – they are seen as responsible for their condition and suffer greater social exclusion [38]. In a response to this, patients may strongly hold onto the identity of being a "normal" drinker, regardless of their use of alcohol. The general unreliability of patient histories has also been previously identified as a barrier to detection [17, 33].

The perceived threat to the patient-doctor relationship is the third major barrier. Attempts to probe and elucidate patient usage of alcohol can be uncomfortable to both the patient and doctor [35], and may be perceived as an attack on the patient's integrity [34]. GPs at times will consciously avoid challenging patient statements on their alcohol use, in order to build and preserve relationships [37]. Our study suggests that this threat is moderated by expectations of the patient-doctor interaction and the perceived acceptable scope of a consultation. These expectations and perceptions are also influenced by sociocultural beliefs.

Although the themes we identified have been described before [28], what emerges particularly strongly in our study is the influence of sociocultural beliefs and attitudes. We propose that its impact on the detection of at-risk drinking in primary care has been substantially underemphasised – recent categorisations of barriers have focussed mainly on clinician factors (e.g., knowledge, time, confidence), some organisational factors (e.g., lack of financial incentives) and occasional patient factors [28]. Interestingly, research conducted in 1984 using a similar qualitative method to this study, found very comparable perceptions in British GPs – the problem of stigma, the difficulties reaching a diagnosis, the emotional aspects of the patient-doctor relationship, and the overarching effect of social factors [35].

There are two important implications. If the enduring barriers to detection are mainly due to sociocultural factors, then this may explain the low uptake of alcohol screening questionnaires, such as the AUDIT-C. This tool was not designed to address these barriers and factors. GPs are not likely to use a screening questionnaire if it is not perceived as useful in consultations. Furthermore, if sociocultural factors better account for the barriers to detection than intrinsic clinician factors (such as knowledge or time), then strategies that focus on clinicians (e.g., training on screening questionnaire use) are not likely to have a lasting effect. There is some evidence that this is the case [25, 27].

Limitations and strengths of this study

This was an exploratory qualitative study designed to develop hypotheses and understand why apparently effective alcohol screening tools are not used in primary care. The study samples the views of a small number of Australian doctors in GP teaching practices in metropolitan Sydney. GPs in teaching practices who participate in alcohol research are likely to be more knowledgeable and confident in the detection and management of at-risk drinking, and are probably not representative of GPs in general. Thus, it is unclear how generalisable the findings are to other primary care settings. Nevertheless, the similarity of the identified themes to previous qualitative research in this field supports this study's applicability [28, 35].

In addition, this study adds contemporary Australian GP perspectives to the qualitative literature. Lastly, it offers an interpretation that may explain the low uptake of alcohol screening questionnaires by GPs.

Direction of future research

The significance of sociocultural factors on the detection of at-risk drinking that emerges from our data was hypothesised three decades ago, "in the long run, medical intervention is influenced by extraneous political and social factors" (page 405) [35], and "responsibility for the failure of early diagnosis and treatment of alcohol misuse in medical settings lies neither with the patient or physician but with the relationship of their roles" (page 427) [39]. This hypothesis deserves further exploration and testing.

Furthermore, there is little evidence that the recommendations promoting GP use of screening questionnaires have made a meaningful impact on routine clinical practice [9, 23-27]. Integrated strategies that acknowledge and respond to the social environment of drinking, encourage the community framing of alcohol use in terms of risks and harms, and foster the legitimacy of alcohol use assessment in primary care to both patients and GPs, might conceivably be more effective. Research into the development of such strategies may be warranted.

Conclusions

The detection of at-risk drinking in Australian primary care appears to be heavily influenced by sociocultural factors. These shape the major barriers to detection – the stigma of alcohol dependence, GP perceptions of unreliable patient alcohol use histories, and the threat to patient-doctor relationships from alcohol use assessment. Screening questionnaires do not appear to address these barriers.

Although detecting at-risk drinking was considered important, alcohol screening questionnaires were not perceived as part of routine practice by our participants. Universal screening was seen as impractical, and the AUDIT-C in particular was considered to have poor practical utility. In the current context, it is unlikely that approaches that focus on GP use of these tools will be successful at improving detection of at-risk drinking in primary care.

Acknowledgements

The authors thank the GPs who agreed to participate in the focus groups and allowed for the interviews to be recorded and analysed. We also particularly acknowledge Natalie Healey for organisational and educational support, Oshana Hermiz for assistance with the focus groups, Melanie Marshall for organisational support, and Joel Rhee for assistance with Nvivo 9.

This research was conducted as part of Tam's study in the Master in Mental Health (General Practice) at the New South Wales Institute of Psychiatry. He was supported by a scholarship provided by the New South Wales Health Department.

Author contributions statements: the research design, data collection, analysis, and initial draft of the manuscript were performed by Tam under the mentorship and guidance of Zwar and Markham. All authors had full access to the data. All authors contributed to the latter drafts, and approved the final version.

Declaration

Funding: nil

Conflict of Interest: None to declare

Tables

Table 1

Characteristics of the participating practices and focus groups

	No. participants	No. doctors in the practice
Practice 1	6	10
Practice 2	7	9
Practice 3	4	6
Practice 4	2	7
All four practices were self-described as "well established", "appointment based" and used "Best Practice Clinical" software for clinical records.		

Table 2

Focus group trigger questions

1	What proportion of adult patients (aged over 18 years) presenting to Australian general practice do you think drink at at-risk levels?
2	What do you think about identifying patients with at-risk/problem drinking in primary care?
3	How do you currently identify these patients?
4	What do you think about routine alcohol screening?
5	What do you think about the AUDIT-C tool?

Table 3

Summary of participant characteristics (self-reported)

	Age (yr) mean (range)	GP experience (yr) mean (range)	Female	Male
GP supervisors	51 (37-73)	22 (11-36)	5	2
GP registrars	35 (30-40)	3 (1-6)	1	3
Other GPs	46 (31-72)	14 (5-26)	3	4
PGPPP [†]	25	n/a	1	
Total			10	9
Other details	4/19 doctors had worked in a drug and alcohol unit. 16/19 doctors were Australian medical graduates. Average hours worked per week: 32 Average patients seen per week: 83 Average appointment length: 16.4 min			

[†] Prevocational General Practice Placement Program – primary care rotations for junior doctors who are employed by teaching hospitals.

References

1.   Britt H, Miller GC, Charles J, Henderson J, Bayram C, Pan Y, et
        al. General practice activity in Australia 2009-10. General
        practice series no. 27. Cat. no. GEP 27. Canberra: Australian
        Institute of Health and Welfare, 2010

2.   Ministerial Council on Drug Strategy. Priority Area 3: Health
        Impacts. In: National Alcohol Strategy 2006-2011. Canberra:
        Commonwealth of Australia, 2006:21-5

3.   Kaner EF, Dickinson HO, Beyer F, Pienaar E, Schlesinger C,
        Campbell F, et al. The effectiveness of brief alcohol
        interventions in primary care settings: a systematic review.
        Drug Alcohol Rev 2009 May;28(3):301-23

4.   Reid AL, Webb GR, Hennrikus D, Fahey PP, Sanson-Fisher RW.
        Detection of patients with high alcohol intake by general
        practitioners. Br Med J (Clin Res Ed) 1986 Sep
        20;293(6549):735-7

5.   Rydon P, Redman S, Sanson-Fisher RW, Reid AL. Detection of
        alcohol-related problems in general practice. J Stud Alcohol
        1992 May;53(3):197-202

6.   Berner MM, Harter M, Kriston L, Lohmann M, Ruf D, Lorenz G,
        et al. Detection and management of alcohol use disorders in
        German primary care influenced by non-clinical factors.
        Alcohol Alcohol 2007 Jul-Aug;42(4):308-16

7.   Yamada K, Maeno T, Waza K, Sato T. Under-diagnosis of alcohol-related problems and depression in a family practice in Japan. Asia Pac Fam Med 2008;7(1):3

8.   Aalto M, Pekuri P, Seppa K. Primary health care professionals' activity in intervening in patients' alcohol drinking: a patient perspective. Drug Alcohol Depend 2002 Mar 1;66(1):39-43

9.   Britt H, Miller GC, Charles J, Henderson J, Bayram C, Valenti L, et al. General practice activity in Australia 2000-01 to 2009-10: 10 year data tables. General practice series no. 28. Cat. no. GEP 28. Canberra: Australian Institute of Health and Welfare, 2010

10.   Harris M, Bennett J, Del Mar C, Fasher M, Foreman L, Furler J, et al. Early detection of problem drinking. In: Guidelines for preventive activities in general practice, 7th ed. South Melbourne: The Royal Australian College of General Practitioners, 2009:37-9

11.   National Institute for Health and Clinical Excellence. Alcohol-use disorders: diagnosis, assessment and management of harmful drinking and alcohol dependence. London: National Institute for Health and Clinical Excellence, 2011

12.   U.S. Preventive Services Task Force. Screening and behavioral counseling interventions in primary care to reduce alcohol misuse: recommendation statement. Rockville: U.S. Preventive Services Task Force, 2004

13.   The Royal Australian College of General Practitioners. Smoking, Nutrition, Alcohol and Physical activity (SNAP), A population health guide to behavioural risk factors in general practice. South Melbourne: The Royal Australian College of General Practitioners, 2004

14.   Berner MM, Kriston L, Bentele M, Harter M. The alcohol use disorders identification test for detecting at-risk drinking: A systematic review and meta-analysis. J Stud Alcohol Drugs 2007 May;68(3):461-73

15.   Reinert DF, Allen JP. The Alcohol Use Disorders Identification Test: An update of research findings. Alcohol Clin Exp Res 2007 Feb;31(2):185-99

16.   Meneses-Gaya Cd, Zuardi AW, Loureiro SR, Crippa JAS. Alcohol Use Disorders Identification Test (AUDIT): an updated systematic review of psychometric properties. Psychol Neurosci 2009;2:83-97

17.   Beich A, Gannik D, Malterud K. Screening and brief intervention for excessive alcohol use: qualitative interview study of the experiences of general practitioners. BMJ 2002 Oct 19;325(7369):870

18.   Brady M, Sibthorpe B, Bailie R, Ball S, Sumnerdodd P. The feasibility and acceptability of introducing brief intervention for alcohol misuse in an urban aboriginal medical service. Drug Alcohol Rev 2002 Dec;21(4):375-80

19.   Bradley KA, DeBenedetti AF, Volk RJ, Williams EC, Frank D, Kivlahan DR. AUDIT-C as a brief screen for alcohol misuse in primary care. Alcohol Clin Exp Res 2007 Jul;31(7):1208-17

20.   Bradley KA, Kivlahan DR, Williams EC. Brief approaches to alcohol screening: practical alternatives for primary care. J Gen Intern Med 2009 Jul;24(7):881-3

21.   Wallace P, Haines A. Use of a Questionnaire in General Practice to Increase the Recognition of Patients with Excessive Alcohol Consumption. Br Med J (Clin Res Ed) 1985;290(6486):1949-53

22.   Goodyear-Smith F, Arroll B, Sullivan S, Elley R, Docherty B, Janes R. Lifestyle screening: development of an acceptable multi-item general practice tool. N Z Med J 2004 Nov 5;117(1205):U1146

23.   Nygaard P, Paschall MJ, Aasland OG, Lund KE. Use and Barriers to Use of Screening and Brief Interventions for Alcohol Problems Among Norwegian General Practitioners. Alcohol Alcohol 2010 Mar-Apr;45(2):207-12

24.   Spandorfer JM, Israel Y, Turner BJ. Primary care physicians' views on screening and management of alcohol abuse: inconsistencies with national guidelines. J Fam Pract 1999 Nov;48(11):899-902

25.   Andreasson S, Hjalmarsson K, Rehnman C. Implementation and dissemination of methods for prevention of alcohol

problems in primary health care: a feasibility study. Alcohol Alcohol 2000 Sep-Oct;35(5):525-30

26. Friedmann PD, McCullough D, Chin MH, Saitz R. Screening and intervention for alcohol problems. A national survey of primary care physicians and psychiatrists. J Gen Intern Med 2000 Feb;15(2):84-91

27. Aalto M, Pekuri P, Seppa K. Primary health care professionals' activity in intervening in patients' alcohol drinking during a 3-year brief intervention implementation project. Drug Alcohol Depend 2003 Jan 24;69(1):9-14

28. Johnson M, Jackson R, Guillaume L, Meier P, Goyder E. Barriers and facilitators to implementing screening and brief intervention for alcohol misuse: a systematic review of qualitative evidence. J Public Health (Oxf) 2011 Sep;33(3):412-21

29. Corbin J, Strauss A. Basics of qualitative research, 3rd ed. California: Sage Publications, Inc., 2008

30. Harris M, Bennett J, Del Mar C, Fasher M, Foreman L, Furler J, et al. AUDIT-C. In: Guidelines for preventive activities in general practice, 7th ed. South Melbourne: The Royal Australian College of General Practitioners, 2009:86-7

31. Ewing JA. Detecting alcoholism. The CAGE questionnaire. JAMA. 1984 Oct 12;252(14):1905-7

32. McWhinney IR, Freeman T. Culture and Context. In: Textbook of Family Medicine, 3rd ed. Cary, NC, USA: Oxford University Press, 2009:122-3

33. Nygaard P, Aasland OG. Barriers to implementing screening and brief interventions in general practice: findings from a qualitative study in Norway. Alcohol Alcohol 2011 Jan-Feb;46(1):52-60

34. Arborelius E, Damstrom Thakker K. Why is it so difficult for general practitioners to discuss alcohol with patients? Fam Pract 1995 Dec;12(4):419-22

35. Thom B, Tellez C. A difficult business: detecting and managing alcohol problems in general practice. Br J Addict 1986 Jun;81(3):405-18

36. Aira M, Kauhanen J, Larivaara P, Rautio P. Factors influencing inquiry about patients' alcohol consumption by primary health care physicians: qualitative semi-structured interview study. Fam Pract 2003 Jun;20(3):270-5

37. Moriarty HJ, Stubbe MH, Chen L, Tester RM, Macdonald LM, Dowell AC, et al. Challenges to alcohol and other drug discussions in the general practice consultation. Fam Pract 2012 Apr;29(2):213-22

38. Schomerus G, Lucht M, Holzinger A, Matschinger H, Carta MG, Angermeyer MC. The stigma of alcohol dependence compared with other mental disorders: a review of population studies. Alcohol Alcohol 2011 Mar-Apr;46(2):105-12

39.   Murphy HB. Hidden barriers to the diagnosis and treatment of alcoholism and other alcohol misuse. J Stud Alcohol 1980 May;41(5):417-28

www.ingramcontent.com/pod-product-compliance
Lightning Source LLC
Chambersburg PA
CBHW030745200526
45160CB00010B/55/J